101

Bladesmithing

Secrets

What Every Bladesmith Should Know Before Making His Next Knife

Wes Sander

Table of Contents

Introduction

In this book, I will share with you 101 secrets of mastering knife making as a hobby or in a professional capacity. These secrets are organized by topic to make it easier for you to learn specific steps in the bladesmithing process. And to improve your reading experience even further, I have arranged the book in a point format. This way, you can easily find out exactly how to go through every step of your newfound journey by referencing this book.

You will be surprised to learn that the entry skills required to make knives are very low. As long as you can wield simple tools like a hammer and sandpaper and operate a belt grinder and anvil, you are good to go! This is one of those skills where you learn on the job. Experience is indeed the best teacher when it comes to making knives. Well, of course at first it will be more like trial and error, but with the help of books like this one, you will become enough of a bladesmith to make knives on a professional (or at least semi-professional) level. And if you look at the bright side, the fact that you learn by trial and error means that your blades will be yours and yours alone.

Free Bonuses for the Readers

To get the most out of this book, I have 3 resources for you that will REALLY kickstart your knife making process and improve the quality of your knives.

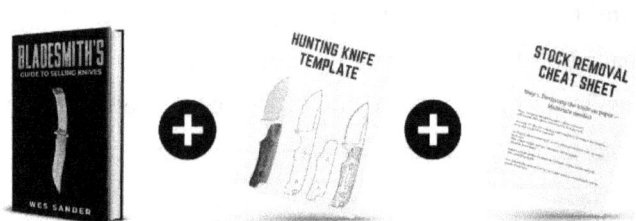

Since you are now a reader of my books, I want to extend a hand, and improve our author-reader relationship, by offering you all 3 of these bonuses for FREE.

All you have to do is go to **https://www.elitebladesmithingmasterclass.co m/free-bonus** and enter the e-mail where you want to receive these resources.

These bonuses will help you:

1. Make more money when selling your knives to customers

2. Save time while knife making

Here's what you receive for FREE:

1. Bladesmith's Guide to Selling Knives

2. Hunting Knife Template

3. Stock Removal Cheat Sheet

Here is a brief description of what you will receive in your inbox:

1. Bladesmith's Guide to Selling Knives

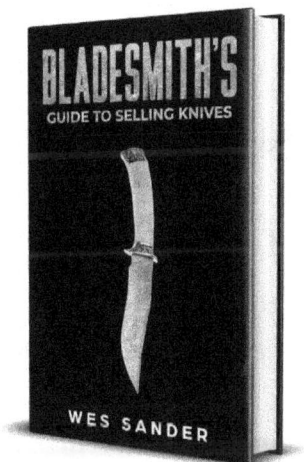

Do you want to sell your knives to support your hobby, but don't know where to start?

Are you afraid to charge more for your knives?

Do you constantly get low-balled on the price of your knives?

'Bladesmith's Guide to Selling Knives' contains simple but fundamental secrets to selling your knives for profit.

Both audio and PDF versions are included.

Inside this book you will discover:

- How to **make more money** when selling knives and swords to customers (Higher prices)

- The **hidden-in-plain-sight** location that is perfect for selling knives (Gun shows)

- Your **biggest 'asset'** that you can leverage to charge higher prices for your knives, and **make an extra \$50 or more** off of selling the same knife.

- 4 critical mistakes you could be making, that are **holding you back from selling your knife for what it's truly worth**

- The ideal number of knives you should bring to a knife show

- 5 online platforms where you can sell your knives

- 9 key details you need to mention when selling your knives online, that will increase the customers you get

2. Hunting Knife Template for Stock Removal

Tired of drawing plans when making a knife?

Not good at CAD or any sort of design software?

Make planning and drawing layouts a 5-second affair, by downloading this classic bowie knife design that you can print and grind on your preferred size of stock steel.

Here's what you get:

- Classic bowie knife design **you can print and paste** on stock steel and start grinding

- Remove the hassle of planning and drawing the knife layout during knife making

- Detailed plans included, <u>to ensure straight and clean grind lines</u>

3. Stock Removal Cheat Sheet

STOCK REMOVAL CHEAT SHEET

Step 1. Designing the knife on paper—
Materials needed

Do you need to quickly lookup the correct knife making steps, while working on a knife in your workshop?

Here's what you get:

- Make your knife through stock removal in just **14 steps**

- <u>Full stock removal process</u>, done with 1084 steel

- **Quick reference guide** you can print and place in your workshop

To get access to this content, go to https://www.elitebladesmithingmasterclass.com/free-bonus and enter the e-mail where you want to receive these 3 resources.

DISCLAIMER: By signing up for the free content you are also agreeing to be added to my bladesmithing e-mail list, to which I send helpful bladesmithing tips and promotional offers.

I would suggest you download these resources before you proceed further, as they are a great supplement for this book, and have the potential to bring an improvement in your results.

Basic guidelines

1. Start slow as a beginner

When you start out as a bladesmith, you'd be best advised to take your time with the making of every blade. With little to no experience, going slow is best in the long run. A slow pace allows you to be meticulous in every step of the knifemaking process and also allows you to learn how to use the numerous knifemaking tools every bladesmith needs. You will be able to avoid mistakes, and your knife will come out looking professional and well done. Keep in mind that bladesmithing is as much an art form as it is a way of making great knives. Every process is an important step in the journey to the final product—a unique and stronger knife you can call your own in every respect.

2. Get the basic tools

Different bladesmiths use a different assortment of tools and equipment. But when you are starting out, you should begin with the following tools: file, sandpaper, hack saw, and drill. While these tools are not optimal and will require upgrading once you advance in bladesmithing, they make just enough of a capital investment that you can enjoy the initial stages of learning how to make knives with little pressure.

File

Files are absolutely indispensable when smoothing the rough edges. You can have files in ranging tooth sizes—from large to very fine—to perform different tasks. You can also recycle your files as blades in the end, so keep that in mind.

Sandpaper

Sandpaper is used to smoothen the blades, and it is used for polishing. Sandpaper comes in various grits.

Hacksaw

A hacksaw is the most basic cutting tool you can use to cut into the metal you use to make a blade.

Drill

A drill is indispensable for boring the holes needed to attach handles, etc.

3. Use the basic tools to make a blade... or two

It is quite possible to make a great blade using the tools described above. You simply get a good piece of steel and use stock removal (stripping off parts of a steel bar to be left with the shape of the knife you want to make) to make a knife. Make a few knives this way, and master this process sufficiently before moving forward with an investment for more expensive tools. Making a knife with these basic tools also greatly boosts your hand coordination, and you'll begin to have a good, steady hand.

By simply adding a hammer, anvil, and a few other forging implements to the four tools mentioned above and making a sandpit fire out in your backyard, you can even advance to the more complex bladesmithing technique of forging. You can then decide the technique that you find most suitable to your specific needs and make an informed decision as to what equipment and tools you want to invest in as you go pro and take your knifemaking to the next level.

4. Stock removal

There are two main techniques of making blades in the knifemaking trade. One is the forging style that requires metalworking skills, considerably expensive equipment including a forge, anvil, hammers, etc. The other bladesmithing technique is called stock removal and entails simply cutting out a sheet of metal until you get the shape of your knife. Stock removal is the easier of the two techniques of bladesmithing. It is the one thing you can do as a beginner without such basic bladesmithing tools as an anvil. Start with stock removal and grow into the art before moving on to bigger things.

As a beginner, you can get knives made using stock removal by outsourcing the parts of the knifemaking process that you can't get done with the four tools outlined above. The process of tempering that requires sophisticated equipment is a prime example. As for shaping the steel into workable sheets of metal, craft stores have stock steel that you can shape into knives. It might take you long because the tools are handheld, but in the end, you get to make knives with very little capital investment on your part.

5. Posture is very important

The way you hold your body while working goes a long way in determining the kind of blades you make. While handling grinding and sanding tools, make sure that you tuck in your elbows. As natural as it is as a working pose, the chicken wing posture gives away the finer controls from your elbows and increases the likelihood of injury.

6. Checking for knife sharpness safely

Of course, you want your knife to be as sharp as it can be. But if you don't know how to check if you have reached the level of sharpness you desire, you risk filing away half of your blade chasing for a sharper edge. So, to be certain just how sharp your knife is, here's what you should look for: a knife that can cut through a paper towel, with minimal pressure. Try to tear a paper towel held loosely from one edge. A knife that can cut through the paper towel without tearing it off is considered sharp.

7. Keep yourself hydrated

Bladesmithing is tiring work. You will find yourself sweating a lot from the fire in your forge or simply performing basic tasks around your smithy. As a result, you lose a lot of body water. The heat from all the power tools means that you might experience spells of dizziness if you don't drink enough water to replace the fluids you lose. An accident in a room full of power tools, metals, open fires, and sharpened blades can be fatal not just to you, but to anyone else in the vicinity. (Fires do often gut down bladesmithing workshops from time to time). You must always stay hydrated by drinking lots of cold water.

Another thing your body loses from sweating because of all the heat and hammering is bodily salts. Replenish these by keeping yourself well fed. This further strengthens your body and makes it able to withstand any strain you put it through.

8. YouTube

YouTube is a great resource for learning knifemaking.

9. You must do it to know how to do it

Experience is hands-down the most important factor in your success as a bladesmith. Bladesmithing is a skill that no amount of book knowledge without practice can teach you. You will gain your most valuable lessons when you apply the theory you learn from books into practice at the forge and anvil. Book knowledge is important as a foundation for every step in the bladesmithing process. Just make sure that you don't get lost reading up on the way to do things that you have no time to actually do anything.

10. Grinder etiquette

A belt grinder plays an important role in the knifemaking process. It helps you to remove unwanted parts of your steel when working on blades. You will use it often, especially in the stock removal processes. Yet the grinder, with its spinning parts, can be a safety hazard as well. Proper etiquette must be observed when working at the grinder.

One thing you must do is ensure that you bring your tool rest as near to the belt as possible to limit movement. Failure to do this increases the danger of getting the knife caught by the belt. What then happens is that the knife can easily get lodged between your fingers, and it could potentially take one off.

11. Apply yourself to the bladesmithing process

Bladesmithing can be a hobby or a professional pursuit, but it is first and foremost an art. From different styles of forging, to materials, tools, and processes involved, making knives requires a special breed of creativity. You should never be scared of expressing yourself through the knives that you make. As much as you must adhere to the standards and processes, it is highly advised that you try one innovative process or design every time you begin making a new knife or sword. The best bladesmiths are those who pushed the boundaries and created novel products by innovating on processes, tools, materials, and designs. Design is especially critical to excellence in knifemaking. If possible, no two knives should be the same unless a customer specifies otherwise. You will be rewarded to the degree that you show commitment and innovation with your knives. So, the more things you attempt, the better it will be for you in the long run.

12. The golden rule of bladesmithing

Every bladesmith makes mistakes. The difference between quitters and those who master the craft is that the latter group is keen to make a misstep and learn from them. The good thing about bladesmithing, especially forging, is that you can correct mistakes simply by heating up your metal once again and hammering. At the same time, there are many ways that you can mess up, and with stock removal, there is slightly less room for error.

But by following the golden rule, no mistake can be too much to dissuade you from the noble profession of making blades. Even if you screw up heat treatment or chip up your first knife with an errant stroke of the hammer, or warp your blade with bad quenching, keep going. Learn from the mistake, record what you did wrong in your journal, and do it right next time. In time, you will master the process and make knives without errors. But for now, embrace imperfections. Let them make you better. Because you **will** get better.

Tools

13. Get a good bench vise

Once you have decided that you like making knives and want to move from the beginner to the intermediate level, invest in a good quality bench vise. It is an absolute lifesaver, and it gives you an outstanding amount of leverage when you are working on your knife. Having your blades held in a fixed position is something you will find very handy. I personally recommend you use a 4-inch vise. The amount of control it gives me is immense, and I can say that the quality of my knife work has improved greatly since I got it.

14. Invest in an air compressor

The 90-degree die grinder and air compressor have the capacity to replace up to 50% of the grinding and shaping tools in your smithy. Even though it will set you back a substantial amount, an air compressor will help you to run power hammers and grinding machines quickly and cost-effectively. This means that you can retire all those grinding and shaping tools that require human force in your smithy and do better work with less effort. Setting it up might be a little complicated, but once you have it going, you will have the power to harness air for functions ranging from the grinding and shaping mentioned above to blowing on your kiln to heat up metals before working on them.

15. Scotch-Brite wheels

The greatest expense you will incur when setting up your smithy will be on abrasion tools. One of the most recommended tools for high-level abrasion is the Scotch-Brite wheel. It pairs well with an angle grinder to give you high-powered abrasion with the added advantage of being relatively cheap. Coming in varying degrees of roughness, Scotch-Brite wheels do the same job as you do with sandpaper, just better, faster and more professionally.

16. Reusing a Farriers rasp

Some of the best blades you can make using stock removal can be from common tools every bladesmith uses. Files and rasps feature prominently on the list of tools you can repurpose to make blades. A used farrier's rasp is especially suitable for new blades because it will save you a whole lot of work when working on suitable handles for your blades. You don't even have to remove the bevels on the rasp completely. The uneven space provides your blade with a good grasp and saves you hours of grinding and sanding to create a smooth surface.

Apart from handles, rasps and files made from metals of appropriate hardness and thickness, like 3/8" brass and aluminum, can be sharpened to make a blade so sharp that it shaves stacked leather like butter.

17. Sanding in tight places

When you start designing blades with complex shapes and curves, you will find that your power tools can't reach them without potentially messing up other sections of your blade. Even the sandpaper cannot reach these nooks and crannies very well when wrapped around your finger. You must be creative here. Wrapping sandpaper in a Popsicle stick or wooden dowel instead should do the trick.

With it, you will be able to access some of the tight corners and achieve a nice finish to your blade. In the same way, you will be better placed to sharpen gut hooks and gouges without too much strain on your part. Depending on the kind of gouge you need to reach, you can use a sharp, curved or oblique dowel to get the job done.

18. New tools

Some of the tools that you will need to use in your knifemaking pursuits will be very expensive. Others like the buffer can be very dangerous, and others like the tempering oven require special skills to use. There is no need to buy a new tool if you won't use it frequently. It is advisable to buy the cheapest tools you can find at first. Test them around the smithy and gauge the quality of their work. Only when you find that you are using a tool a lot should you buy a quality one. For the more expensive tools that you only use on occasion and for short periods of time, outsourcing is also a valid option. That is why, if you cannot afford equipment to heat treat your knives, you should send them out to professionals.

19. The tools you buy

When you become more experienced with bladesmithing and you decide to actually buy a tool because you are using it enough, there is one rule that you must follow; don't skimp on quality. This is especially true for sanding belts and files. They are critical to the knifemaking process and nothing than the best of the best will do. Nicholson is the brand I recommend for files. As for grinder belts, always go for ceramic ones.

20. The right way to hammer

Proper handling of a hammer is to ensure that you always grip the handle with your thumb going all the way around rather than along the length. Not only does this way of handling your hammer give you greater control and grip, but it is also a safety violation to keep your finger along the handle. It makes it very easy for you to strain your radial tendon when hitting metal.

21. Get the right tongs

Tongs are some of the more expensive tools in the smithy. They also come in many different variations. Their usefulness, however, cannot be underestimated. If you can afford to, it is always advisable that you use the right tongs for every task during the forging process. Apart from making it much easier for you to hold hot metal, the quality of your forging will also improve substantially.

Safety

22. Safety rule #1

Your continued wellbeing is your number one priority. Keep the props to the bare minimum at all times. If you don't NEED it, don't take it with you to the workshop. This means that things like phones, keys, etc. should be left outside. On your person, things like rings, bracelets, or necklaces also have no place in the smithy. They just increase the chances of injury, especially if they were to accidentally fall into the machinery or catch on one.

23. Basic safety equipment

The bladesmithing profession entails constant danger from the tools you use, the processes of making blades, and the blades themselves. You must always put on your safety equipment every time you work in the smithy. A respirator is among the most important safety equipment for the bladesmith. It comes in handy, especially while you're sanding and grinding as it prevents the particles from reaching your respiratory system.

Generally, you are supposed to ensure that you keep yourself safe from any risk of harm. If you have long hair, then you must ensure that you keep it tied up. A long beard should be kept away from spinning tools.

Essentially, anything that could cause complications or lead to harm needs to be kept away. For example, smoking in the smithy is a huge no-no. Anything that brings even the remotest chance of danger should strictly be off-limits.

24. No gloves while operating spinning tools

Gloves make up part of basic safety, but they could turn out to be very dangerous too. While operating spinning tools, it is very easy for gloves to get caught and bring harm to the wearer. While we are talking about gloves, ensure that the ones you have are a good fit. Oversized gloves cause slippage and increase the chances of accidents.

25. Safety first while drilling holes

There are two reasons why you should clamp down your knife tightly when drilling the pin holes. The first reason is that you get a more accurate drill hole. The second and most important reason is to make sure that the drill bit doesn't catch the blade and turn it into "the windmill of death" that could very easily bring fatal injury to you. A good many bladesmiths have lost fingers after overlooking the importance of clamping a knife down while drilling. A knife that is rotating at the speed of a drill, whether sharp or dull, is lethal when it comes into contact with human flesh.

26. Caution with machine buffing

A buffer is very effective in polishing blades, but it is by far the most dangerous tool in the bladesmithing profession. In fact, a lot of knifemakers call buffers widow-makers because they have filled bladesmithing forums with stories of fatal accidents. This is because buffers tend to very often catch a knife and fling it away with immense force and speed. If a buffer-hurled knife catches you wrong, it could penetrate muscle to reach veins and bone. **You must be very careful with them.** Start by wearing protective clothing and operating it from a 90-degree angle with the knife pointing away from you.

27. Fire safety

A forge is a constant fire hazard not just to you while you work inside the workshop, but to the house and neighboring houses if you do it in a densely populated area. If possible, locate your workshop in a stand-alone structure well away from other buildings.

More importantly, you must always have a working fire extinguisher within easy reach in your workshop. Train yourself on basic fire response and extinguish any fires before they spread. As for the forge itself, ensure that your coals are dead any time you leave the workshop unattended, even for short periods of time. This ensures that they do not spark up and cause a fire.

28. Belt sander safety

The belt sander can be used to sand both wood and metal, but switching from one to the other can create a fire hazard if it is not cleaned properly. Ensure that you wipe out all wood or metal particles before switching to the other material. And while wiping, use a vinyl cloth or brush, not the bare hand, to avoid accidents.

29. Don't ever zone out

In pretty much any other job, repeated performance of a task is equivalent to mastery and the license to dissociate and let reflexes take over. With bladesmithing, you must pay attention to every single detail of every single stroke of the hammer, immersion in oil, shaping at the grinder, and pretty much everything else you do. You should never do anything on autopilot as this is inviting disaster and mediocre results.

30. Workspace organization

Safety concerns in the bladesmithing trade come not just from the tools and processes, but also the general organization of the workspace. The best way to design your workspace is by using the 5s methodology. This ensures that you don't have to make too much movement between the different machines. It is also advisable to keep all the tools organized in designated areas so that you don't ever have to search for any tool when you need it urgently.

The 5s methodology was created by the Japanese to help with workplace efficiency. It follows the following five principles.

Sort

Sort tools into different marked categories. Keep the tools you use more in more accessible positions.

Set in order

Organize every tool and all equipment for quick and easy location, and label everything.

Shine

Cleanliness is important. Clean your workshop from time to time and degrease every work surface. Cleaner workspaces are safer and pose less danger of harm or injury.

Standardize

Create a standard work procedure that takes you through principles 1 to 3 above. For example, every time you finish making a knife, you sort, set in order and clean everything.

Sustain

Make sure that the four steps above are adhered to no matter what. Turn workspace organization into a habit and you will never regret it.

31. Constant maintenance of equipment is crucial

Maintenance will be needed for every tool in your tool rack and every piece of equipment in your workshop. Maintain your tools and equipment when they are working well and especially when they are malfunctioning. Failure to ensure constant maintenance will increase the chances of a tool or equipment failing at the most inopportune time.

Steels

32. Learn how to work the easier steels first

Bladesmiths use three primary types of steel in making knives. By level of complexity to work them into knives, we have the 1084, the 1085, and the 01 respectively. The O1 is especially suitable for knifemaking because it is longer lasting, easier to maintain as a knife, and relatively inexpensive. It also keeps the edge for a longer time.

However, it is recommended that you focus on mastering the 1084 steel first, even though knives made from 1084 steel tend to rust easily. Make sure that you practice forging and stock removal well while working with 1084. It is much easier to do these things on the more forgiving 1084, meaning that your chances of failure, frustration, and disenchantment with bladesmithing are less with it. You should master the techniques of bladesmithing on 1084 steel and make at least two good knives with it before moving on to the 1095 and O1 types.

33. Rust-insulating your blades

Rust destroys many carbon steel tools and is the reason why they require constant care to keep in optimum performance levels. Stainless steel has a high chromium content in its composition, sacrificing hardness and ability to maintain an edge with the promise of freedom from rust. Blades made in the smithy, however, favor higher carbon content, hardening them and making them sharper for longer than their stainless steel counterparts. The downside to this is that these blades rust much more easily, making them high maintenance and often unsustainable.

I mean, it is hard enough to remember to clean and oil your sheathed blades every once in a while, but what about your forged kitchen knives that you use every day? A mistake like leaving it in a moist environment overnight could be catastrophic. You end up with rusty knives, and if you are not careful, food that tastes like rust as well. Luckily, you can avoid developing a coat of rust on your carbon steel blades by developing a layer of patina on them. The patina is a protective layer of inert rust that coats the blade and prevents further corrosive action. Your blades remain rust-free and low maintenance as you can leave them lying around in humid conditions without spoiling them.

34. Evapo-Rust for restoring rusty blades

Unlike blades made from stainless steel, the knives forged from iron tend to rust easily. This makes it important that you understand how to restore rusty blades, which, luckily, can be done simply by removing the rust off the blade. The best way to do this is to use Evapo-Rust. This non-toxic fluid makes the process of restoring rusty blades a walk in the park as all you need to do is soak your rusted blade for a few hours. You don't even have to be present for the duration; you can run errands or continue working on your new blades as the Evapo-Rust works its magic. A few hours and a good wash later, the rust is gone and your blade is restored to its old glorious state. The same breezy process applies for when you are restoring rusty old tools like hammers. You won't even have to scrub or sand anything. And the best part is that Evapo-Rust comes ready to use. You don't have to mix it with anything; you simply dip your rusty tools in there and everything is taken care of.

35. Rust bluing

If a blade has already rusted and you are looking to clean it, there is a better process than simply removing the rust layer with Evapo-Rust. You see, cleaning using Evapo-Rust is effective in removing rust from your blades, but it just restores the metal to a state where rust development can start all over again. You can expect to have to do the whole process all over again. However, by using a process called rust bluing you can instead convert the rust to the protective patina mentioned above.

Rust bluing is essentially an oxidation process that converts red rust (chemical formula Fe_2O_3) to blue rust (chemical formula Fe_3O_4). The process follows the steps described below.

You will need:

Hydrochloric acid or any other highly reactive acid like nitric acid

Thick, water-resistant gloves

Water in a container

Fire source

Scrubbing brush/denim cloth

Sandpaper

Process:

- Clean any rust or residue from your blade using the scrubbing brush

- Coat with an even layer of acid

- Let it sit for up to six hours, developing a good layer of red rust

- Immerse the blade(s) in the jar of boiling water. The water reacts with the red, losing its oxygen molecules to convert it into the more oxygen-rich and stable blue rust

- Lightly scrub the blade with the denim cloth to remove the excess acid and rust. Be careful not to wash off the patina along with the red rust in this process

- To remove the dull blueness of the patina and make your blade shine, dry the patina out for about twelve hours and then sandpaper evenly

36. Alternative rust bluing method

If you do not have the materials to do the oxidation process as described above, you can follow the easier method of simply dipping the blade in a jar of boiling distilled water with no prior treatment. The hot water speeds up the rusting process, but instead of the red rust, you will see a dark patina forming on the blade. This patina will protect your blade from rusting just fine. But a word of caution, sometimes you will boil your blades in hot water for hours with no blue patina forming. This is normal; sometimes distilled water simply doesn't work. If this happens, just try a different water source. I have had blades that only formed a patina when dipped in water from my A/C unit. Experiment with different waters until you find the one that works with the type of steel you have used.

37. Procuring your steel

Local stores are unreliable for procuring knifemaking steel. Instead of relying on them for your steel, go directly to the suppliers. Stores like New Jersey Steel Baron, Beltway and Metal Supermarkets are also great, especially for bulk purchases.

38. Working with stainless steel

Sometimes, you will find it necessary to use steel with the same chemical composition as stainless steel, such as stainless cable. While it might make considerably good knives after hardening and quenching, you must be very careful when forging stainless steel. The galvanization process that makes it resistant to rust infuses it with the element chromium, which releases fumes at the forge that have the potential to make you very sick, especially after a long period of exposure.

39. Making a Hamon blade

Hamon blades are a type of Japanese swords that have a wavy line on the edge. The process of making a Hamon knife is similar to that of making a regular knife, but with a slight variation. When it comes to heat treatment, you use the same clay coating that is used in katana swords and apply unevenly on the spine. The edge of your knife should receive a thinner coat so that it hardens more than the rest of the spine and keep a sharp edge for longer.

But, more importantly, if you are making a knife with a Hamon, remember that etching is the only way to bring out the Hamon. The citric acid found in lemons is one of the most efficient liquids you can use to really bring out the Hamon line.

40. Jackhammers

You can make some terrific tools with Jackhammer tips. Since they are usually made from the 4140 type of steel that carries high impact toughness, they are great as drift punches and hammers. While you can also make some fair knives from jackhammer tips, the fact that 4140 does not harden well limits the applications of such a knife.

Heat treatment

41. Hardening

Steel only gains utility after it has been hardened. The process of hardening steel entails heating it to high temperatures, then quenching it in oil. The rapid cooling strengthens the bonds between iron and carbon atoms and makes the steel tougher to break. One thing to keep in mind while hardening steel is that the tips normally harden at a faster rate. You must, therefore, take care to avoid over hardening them.

42. Making a quench tank

If you are looking to start making knives by forging them, a quench tank in which to cool the metals as you work on hardening them is very important. A cheap way to make one is to procure a tin of canola oil. Buy simply cutting the top off, you get an instant quench tank. Canola oil is especially suited for quenching 1084 steel. When quenching 1095, it is more advisable that you use Parks 50 instead. Most bladesmiths use motor oil in their quenches, and while this is all fine, there are certain precautions that you need to take to protect yourself from possible harmful fumes. Moreover, using oils other than the ones named above means that you can never quite get the very best levels of hardening in your steel.

43. Maintaining your quenching oil

Quenching is the process through which heated steel is rapidly cooled by dipping in oil. The kind of quenching oil you use will play a big part in the quality of hardened steel you get in the end. It also impacts your safety as the quencher (more on that below). But, most importantly, the way you maintain your quenching oil has a huge impact on its pace of deterioration. With every quench that you do, deposits of steel are left in the quench tank and often react with other impurities to produce clogs inside the oil. If you maintain your oil properly, it could remain useful for a very long time. Failure to practice proper maintenance will always result in a clogged quenching tank and improperly hardened steel.

To take care of all these challenges, I recommend occasionally heating up your quenching oil a few hours before you start quenching your steel. The heat helps break up the oil particles and makes the quenching process faster and more effective and your oil longer lasting and less poisonous.

44. Quenching oils to keep away from

Not every type of oil can safely be used for quenching. Some of the commonly available oils, like motor oil, produce very toxic fumes. You must practice due care while using motor oil in the quenching process. While its thicker structure allows for better absorption of heat and hardening, the fumes that are released from motor oil can be incredibly harmful, capable of inducing dizzy spells and headaches.

45. Fixture tempering

The heating and cooling process that the steel undergoes during quenching might leave it with warps that make it unsuitable for making blades. Knives made from this kind of steel might turn out to be bent and of poor quality. You must remove these warps before proceeding to the final knifemaking stages.

By far the most effective way to get rid of warps after the hardening process is fixture tempering. With fixture tempering, the steel is tempered under pressure, between two flattened sheets of metal, for example. Fixture tempering is more effective than re-hardening the steel because it guarantees results in a fast and relatively stress-free way. The steel also tempers uniformly, unlike re-hardening that might produce steel with the old uneven levels of hardness and results in the warps all over again.

46. How to test for proper hardening

No process is complete until you can be sufficiently satisfied, beyond a shred of doubt, that you have achieved your intended result. Yet with hardening, it is very hard to know how well the hardening process worked short of taking a hammer to the steel. This, however, is out of the question because it will destroy your work up till this point. So how exactly do you find out if your piece of steel has reached the desired level of hardness?

Well, you can try running a file over the blade and slanting it slightly. If the file glides along, then you can be satisfied that the steel has achieved an acceptable level of hardness. If the file just sits on the blade, it is an indication that you have to restart the heat treatment all over again.

Handles

47. Basic handle-making guidelines

While inserting the pins into the handle, you must make sure that they slide in smoothly. To do this, ensure that the pin is appropriately sized to align with the holes you drilled. This way, you stand with a lesser chance of blowing out the handle material by warping the pins with forceful hammering.

48. Hand sanding

To get a good polish on your blade, you need to hand sand. Sandpaper comes in handy with this but tends to get clogged with the particles scrubbed off. You can take care of this by simply using a window cleaner, which will make it easier and keep the paper from getting clogged easily.

It also breaks the sandpaper down and ensures that your polish comes off smooth and even with minimal scratch marks. The ammonia is very good for bringing out the grain lines in exotic woods like maples, olive, etc.

49. Special knife handles

A knife's handle does not always have to be generic and boring. You can try different designs to give it more character. These designs include hidden tangs, guards, spacers, and bolsters. With the hidden tang, you make a handle that swallows up the part of your blade that falls under the handle (tang). A guard is usually a piece of metal that covers the front part of the handle, separating it from the blade. Spacers help you to break the monotony of a single handle material and bring a unique, distinctive look to the handle. Finally, a bolster is a band that extends from the blade and forms a junction with the handle. It creates a seamless transition from blade to handle and allows for balance between the blade and the handle.

Now all these special handle designs require special sanding techniques. For example, when sanding a handle with a guard or natural and metal spacers, or smoothing the joint between blade and handle in knives with bolsters, you must use something as a hard backing when sanding. This ensures that you sand everything evenly and avoids the accidental over-sanding of the softer bolster material. With hidden tang blades, ensure that the groove you leave for the tang is just big enough to avoid forcing it in.

50. Making great handles

A good knife is as much the material that makes the cutting edge as the handle that allows the blade to be brandished. Micarta is one of the best handle materials you will find out there. The thermosetting process used in the making of Micarta means that you get a durable, multi-layered handle that maintains a great grip even when wet. A great way to finish off your Micarta handles with a strong bond is to apply Cyanoacrylate glue (CA). CA glue works even better with Micarta in custom-made blades because it allows the material to reach a higher level of shine than other handle materials. This is achieved by rubbing the handle with fine steel wool in between coats of CA glue and then buffing.

51. Drilling holes in the steel

Other than decorative niches punched through a blade, holes are usually drilled to fit handles. The first rule when drilling holes is to make sure that you always do it BEFORE heat treatment. The steel is relatively soft at this time, but once it has been hardened, it becomes tougher to drill holes in it. Doing it after heat treatment means that you will encounter very tough steel and risk ruining your drill bits.

52. Alternative ways to drill holes

If you are struggling with drilling holes into your blade, it might help to try to drill into the knife design that you sketch on the steel bar before it has been cut out. You will discover that it is much easier to hold a rectangle steel bar in a clamp, than a knife blank. This technique is especially useful if you are using stock removal rather than forging in the making of your knives.

53. Drilling holes into wood

Wood is one of the most common materials for making blades to be used in forged knives. Bladesmiths use everything from matured, burly, and knotty wood stock to commonplace timber pieces from the local lumber yard. Whatever your source of wood, you don't want to destroy it while driving pins through them. To avoid this, you must do the following:

- Fasten the piece of wood that you intend to make into a blade tightly against another, less valuable, sacrificial block of wood.

- Drill your holes through the wood and pierce into the wood down below. The drill bit is usually thinner at the front, so going straight through ensures that the hole is drilled evenly.

- When the pin is finally driven into the handle, it will go straight through like the drill bit and a blowout will be avoided.

54. Use brad-point drill bits

Alternatively, you can use brad-point drill bits to position your hole even more accurately. Brad-point bits are specialized drill bits for drilling through wood because they facilitate the precise location of holes and perforation of exact-sized holes. The brad-point bit is specially designed with a nail-sharp point that allows it to bite into the wood and start drilling with pinpoint accuracy. Not only does the brad-point drill ensure that the hole you drill is evenly wide, it perforates a hole in accordance with the grain of the wood for greater overall strength of the handle.

55. Leather stacked handles

Leather is one of the best materials for jungle knives or other blades that require a versatile handle. It retains its handiness wet or dry, is much lighter than wood and metal, and lasts a very long time. Stacked leather is even better because it creates an exquisite finish to your knife and is relatively easier to shape. The following procedure will get you a kickass leather-stacked handle for your knife.

- First, cut your leather into squares. These squares are supposed to be oversized to give room for shaping later on.

- Punch a hole in the middle of the leather pieces. Because leather expands, the hole can be slightly smaller. This allows for a tight fit. If you want, you can mix it up with other materials like epoxy.

I have found that this creates an even more interesting blade design in the end.

- With all the pieces cut and fit, soak them all in glue and begin placing them through the tang.

- Apply pressure (for example, by screwing in the pommel) to create a tight fit.

- Give it a few hours to harden and then cut out the extremities. You can do the final shaping by simply sanding with rough grit sandpaper and moving on to finer ones.

- With the shape achieved, it is now time to polish. I have found that Karnuba wax works best on leather, giving you a smooth, shiny surface with all the colors of your leather pieces highlighted.

56. Removing epoxy

You should always cover the blade while attaching the blade to the tang using epoxy. But in the event that you mess up and the epoxy clings on to the bolster or the blade, don't panic, just apply acetone and Q Tips, then use a sharpened brass piece to chisel off whatever is left. You can then sand and polish to regain the old shine of your blade.

57. Painting on leather sheaths

The sheath is one more thing you can use to make your knife even more outstanding. When you make it out of leather, you can paint any number of patterns or shapes you want. As a rule, we make a coating of white paint on all the areas that will contain artwork. Let the sheath dry out completely and then put your preferred color on top of the white. This way the white undercoat brings out the true color of your paint and makes the color pop. Moreover, it prevents the dye color of the sheath from trying to push through underneath your painted color, which makes the leather sheath appear brownish if left unchecked.

Another secret to painting on leather is that you must use specialized paints, which work best with absorbent leather surfaces. Finish the work off with some varnish for the slightest glint and you have a sheath that does your knife proper justice.

58. Creating a sheath with different types of leather

There are quite a few options available for you to choose from when deciding what type of leather to use with your knife. The two main types of leather used with knife sheaths include upholstered and veg tan leather. The way you work with upholstery leather is very different from working with veg tan leather.

For one thing, when you buy upholstery leather, it often comes dyed and sealed, meaning that you don't have to wet it beforehand. On the other hand, you must wet veg tan leather, which also comes with no dying or sealing, before starting your work on it. The wetting of veg tan leather ensures that you don't tear it when tightening the joint stitches, which is not a problem with the much tougher upholstery leather. And while we are on the topic, it is much better to make your sheaths from high-quality upholstery leather rather than veg tan leather, especially when your knife is of quality design. A good sheath goes hand in hand with the blade and fetches the knife a better price in the market.

59. Treating your leather

Leather deteriorates in quality when let to sit in humid conditions for a long time. To dry your leather after wetting, you can speed dry it in low oven heat or let it out under direct sunlight. Otherwise, the wet leather will become a haven for bacteria that eat into the fleece and turn it funky.

Wet leather sheaths also increase the chances of your blade rusting. Simply storing your blade in a leather sheath increases the chances of rusting because leather attracts and retains moisture. Having the leather wet or in any way humid just exacerbates the situation.

Another way to make sure that you don't spoil your leather while wetting it is to make it workable is to spike the water you use with isopropyl alcohol. This reduces the boiling point of water and ensures that it dries much faster and slows down the growth of bacteria on the leather.

60. Sanding the handle

The handle is one of the most important parts of a knife, mostly because it affects the way we wield it. It is very important, therefore, that you make it as smooth as humanly possible. To do this, you must use a progression of belts to sand the handle. Start with 60 grit and take it to 80, 120, and then 220.

The scales are supposed to fit perfectly with the tang without leaving any awkward spaces. This is achieved by carefully sanding the backs of both scales flat before attaching them to the tang. The tang must fit in perfectly with the scales with no edges showing or be completely covered. If you realize that the edge of the tang is visible, you can either get new scales of appropriate size.

61. Gluing the scales to the tang

The scales are supposed to fit perfectly on the tang and remain attached for as long as the knife will be used. A tight fit is thus needed, which can be achieved with epoxy glue. However, before applying the epoxy, you should rub the inner part of the scales where the epoxy will be applied with acetone. This removes any traces of moisture in the scales and ensures that the parts stick together even better.

Foldable knives

62. Basics of folding knives

The most important mechanisms of a foldable knife are present inside the handle. This is not to say that the blade is just like that of a common knife either. Not at all; there are nuances involved in making a foldable knife blade.

The main parts of a folding knife are the pivot, the lock, the stop pin, and the flipper.

The pivot plays the dual role of attaching the blade and the tang and allowing the blade to flip out from inside the handle. The lock keeps the blade from turning back into the handle and keeps it open for use. A titanium spring is the most commonly used lock mechanism. The stop pin is the part of the handle that extends over the pivot to press the blade down and allow for cutting. Without a stop-pin, the blade would keep pivoting to lie on the other side of the handle.

In simpler models, the flipper may simply be a slight grove on the blade where you can put your finger and turn the blade out. But in the more sophisticated types of knives, a quick-release mechanism allows you to release the blade with a single press.

The handle of a folding knife has to be grooved to allow for the blade to turn inside, which means that the tang and blade come in two separate steel pieces joined together on the pivot.

63. Making a pivot that works for your folding knife

A folding knife is more complicated to make than other knives because it has a rotating blade. It only has utility when the blade turns the whole way from the groove in the handle and is held by the stop pin. To complete the turn properly, the pivot needs to be smoothened around the edges. This ensures a swift and friction-free turning radius and avoids wear and tear. While you make your foldable knife, be sure to use a drill to turn the outside of your pivot ends. The drilling must be precise, so try to use a drill guide. You can also insert a screw into the pivot and turn it a few laps at slow speed to check if the pivot is in good working condition.

64. Shortening the pivot

Sometimes you realize that the pivot you have is too long and have to rework it to make it shorter. To do this, you should make a test piece of stock 3 reamed pivot about a quarter inch in length. Make it complete with a stop pin sized hole as if it will be used in your knife. With a belt sander, square the end of the pivot until you reach the requisite shape and size.

65. The stop pin

The stop pin ensures that the blade does not close into your hand when you are using the knife and hurt you. To ensure that your stop pin works as it is supposed to, you must make sure that you assemble the different parts of the knife properly. A good procedure to follow is to start by assembling one side of the knife, then using a super fine Sharpie to mark the point where you want your stop pin.

You then hold the blade firmly against the tool rest and use the sander to smooth it through till you expose the line. Alternatively, you can hold it against a 60 grit drill and shave it off until the mark shows up. With the super fine Sharpie, the mark shows up as a solid line when spinning. You can then switch to a 220 grit-size belt and sand it slowly until the mark is gone. Follow it up with squaring and your blade is good to go.

66. Tempering the titanium spring

When making a foldable knife, I have found that tempering the spring to 50 HRC makes it less prone to future breakages. Your knife will keep springing back and forth into the groove long into the future.

67. The perfect finish

To get the perfect finish on your foldable knives, apply a mirror polish to every component of the knife before assembling them. From the pivot, the blade, to the barrel, all the components work better when they are highly polished. With proper waxing and/or oiling, the components will remain resistant to deterioration from friction to a great degree.

Grinding

68. Choosing a grinder

The following are the most important characteristics to consider when choosing a grinder to buy:

The frame

The heavier the frame, the better the grinder functions. This is because the bulk and weight of the grinder absorbs most of the vibrations and makes it easier for you to work.

The motor

Find the best possible quality motor. The maker is very important, so look for quality-branded motors. They last longer than generic ones.

Variable speeds

Multi-speed grinders come with a higher price tag but give you the option of choosing slow or faster speeds to grind on. This comes in very handy, especially for the beginner. If you can afford it, then by all means buy a multi-speed grinder.

Versatility

Some grinders come with several attachments. These attachments are often replaceable and can come in very handy when shaping various parts of your knife. Go for it if you can afford it.

69. Grinding angles

A good angle for the edge is about 20-degrees and about an eighth of an inch in width.

70. Proper use of the grinders

Grinding is relatively hard work that will drain your body of energy and put a lot of strain on you. You should always maintain a comfortable posture while grinding to prevent back problems. This means that your back should be straight, not stooped, and you shouldn't have to lean down or lift your knife too high to reach the grinding belt. Proper use of the grinder, therefore, starts with the workbench on which you place it. Customize your bench with a focus on your height.

Running your knife through the grinder belt should be done carefully, with constant attention to grind lines against the knife template you made. A firm grasp is a must. When using the angle grinder ensure that your blade is clamped down tightly with a bench vise to ensure precision and avoid possible injury from the grinder sending the blade flying. Alternatively, you can clamp the steel using C and G clamps.

71. Grinding technique

To get the best grind lines, follow the following guidelines:

- Your hips give better control for shifting grind lines. Use them instead of your wrists.

- The wider you hold your elbows, the less control you have over the grinding process. Keep them tucked close to your sides and use the leverage against your ribs to get the best possible grind lines.

- Let your shoulders relax back into its socket and keep your stomach tight.

Not only will you get more control of your grinding this way, but you can work for longer without getting tired.

72. A light touch works best

Maintain your confidence and cool around the grinder. At no time should your knife ever be pressed hard against the belt. The only thing you achieve by doing that is absorbing the vibrations from the belt into your hand, increasing the chances of a mishap that could ruin your grind lines. Unless you know how to use it, the machine will always win. If you don't use it properly, you will end up with a poorly done knife simply because your grinder game is not up to the mark.

73. Grinding curved blades

A curved blade makes for a great design but isn't always the easiest to grind. Even with an angle grinder, a curved blade is a hassle to get right. With the belt grinder, you can make it possible to work on your curved blades by simply shifting the belt off to the side edge. Because the edge is sharper than the belt groove, the belt forms a sharper section that you can take advantage of to grind curved blades.

Figure: Knife with tasteful curves.

74. Different types of grinds

The grind on a knife determines the use to which it can be put, its cutting strength, and its balance.

- A **hollow grind** gives a concave cutting edge that improves slicing quite dramatically, even though it has been found to make a knife's edge rather weak. A knife with a hollow grind will have a razor-sharp edge. In the making of a hollow grind, try not to worry too much whether the grinding belt wanders or not. After you have made a groove, the belt tends to follow the groove of the hollow grind on the blade. This does not mean that you should be any less careful, but making that first groove does make it easier to hollow out your blade.

- A **flat grind** tapers consistently from where the cutting edge starts to the tip. There is no curve to the grind, which means that the cutting edge could extend all through to the spine, as is the case in a full flat grind. The other two types of flat grinds; the high flat and Scandinavian grind, both start midway.

- A **chisel grind** is another common grind that has only one side of the blade sharpened. The side that is not sharpened is completely flat while the other has a sharp edge starting from the middle. The chisel grind is among the sharpest there is, but it keeps an edge for a shorter period of time than the double-beveled grinds.

- The **convex grind** curves outward to form a convex edge. Because of the convex nature of the grind, sharpening is considerably harder and the edge is not as sharp as other grind types. However, the convex grind keeps its sharpness for much longer. It is a specialized edge that is found in chopping rather than cutting blades. A heavy blade helps to compensate for the lack of cutting power with power.

- A **compound bevel grind** uses two convex grinds, the cutting one being smaller and tapering to a sharper edge. Compound bevel grinds deliver cutting edges that are as good as flat and hollows in sharpness without compromising much on the durability of the blade.

- Finally, we have the **asymmetrical grind**, a bevel that has a convex grind on one side and a concave grind on the other. The asymmetrical grind also improves sharpness (it has a razor-sharp edge) and reduces the possibility of chipping.

The choice of grind for your knife must align with the use for which the knife is intended.

75. Knife tips

As stated above, you should make every knife with the end use in mind. And as far as tips go, the rule is that a knife will only cut as well as the angle of its tip. For example, if you want to use your knife to slice through soft things like meat and vegetables, a thinner tip is the way to go. In the same way, make a knife with a thick tip if you intend to use it to cut through harder materials like wood. The tip completes the grind of your knife into razor sharpness or handy chopping power.

76. Hack for getting the best grind lines on your blade

If you want to get a really good grind line on your blade and you don't mind working for it, you can draw a top line for the bevel on both sides of the blade and a center line where you want them to meet to form an edge. Work within these three lines with stone wheels and then switch to files to bring the bevel shape. Just keep the type of grind you want in mind and either gouge a concave shape or leave a convex swell to get your desired grind type.

From experience, I can tell you that the compound bevel is the most complicated type of grind because it requires immense hand control. The chisel and flat grinds are the easiest.

77. Making bevels using a grinding jig

It can be a hassle to make proper bevel lines in your blade. When this is the case, a grinding jig can be very handy. You see, the biggest problem with making bevels is keeping them even throughout. A grinding jig solves this issue by moving your knife right along with the grinder.

Building a grinding jig requires:

Materials:

A door hinge

A five-inch threaded bolt

A nut that fits around the bolt

Two quarter-inch thick steel plate measuring about six inches by two inches

A similar-sized plastic sheet

A block of steel of dimensions four by two by three quarters inches

More bolts of varying sizes

Process

- Drill holes into the two steel plates and attach the hinge with appropriately sized bolts.

- Bore a hole into the steel block and attach it to one of the steel plates

- Bore another hole through the side and attach it to the other plate using the threaded bolt. The angle of the hinge can be adjusted by simply twisting the bolt.

- Drive a pin into the hinge to tighten it up.

- Bore a hole into the block of steel to touch with the bolt within. Drive a screw into this hole and use it to tighten the bolt and prevent the jig from any kind of movement.

- When grinding, you simply attach the knife to the jig at your preferred angle and bring it close to the grinder. Moving the knife by moving the jig ensures an even, professional grind.

78. Working with the grinder at different speeds

The grinder comes in handy in many processes and can be mounted with diffcrent kinds of wheels to perform various tasks. When grinding, the slower speeds can be a great advantage. A slower speed allows you to achieve higher levels of precision. However, you must learn the proper way to lay the blade on the wheel with such a light touch as to remove no matter at all. This way, you can do some of the sandpaper fine sanding that takes hours upon hours to do by hand and save yourself a lot of energy.

Alternatively, you can set the speed of your grinder into low and use it as usual, but remove only the smallest pieces from your blade. This is why a variable-speed grinder can be a great investment for your workshop. One with a knob to change speeds is especially useful because you can set it at the lowest speeds and sand your blade without worrying about breaking it.

79. Serrated blades

The serrated blade is one of the most dynamic edges for knives, especially for professional kitchen use. Knowing how to make one can be very important to serve your clients' needs. To create serrations on a knife, you start by marking even divisions on the knife using a caliper and dykem blue fluid.

With the blade clamped firmly to a vise, use a Nicholson chainsaw file to file it off on one side, leaving the other to be the sharp edge. When it comes to sharpening, you do it from the non-filed side only. It is most advisable to file the blade before proceeding to the heat treatment step, leaving an edge of about 10-15 thousandths of an inch in thickness.

80. Cleaning your sharpening stone

It is important to keep your sharpening tools in good working condition because this affects the sharpness you can get on your knives. Failure to clean the sharpening stone will result in metal filings embedding on the surface and turning your attempts at polishing your blades into polishing sessions where your blade comes out sparkling but no sharper than it was before. There are three simple ways to clean your sharpening stone.

The first is to pour a small amount of honing oil on the stone and rubbing it in with your finger or a toothbrush. The oil digs out the metal flecks and makes it possible for you to wipe them off with a toothbrush or using a damp cloth. You can then run the stone under running water to restore it to its optimal working condition. DO NOT use soap while washing a sharpening stone.

WD-40 is another thing you can use to remove dirt from your sharpening stone very easily. Simply spray it onto the sharpening stone in a well-ventilated area, scour the stone with fine steel, then wipe off any grime with a paper towel. WD-40 is especially effective for removing grease and oil.

Flattening is the last method of maintaining your sharpening stone in good working condition. A flattening plate helps you to even out the surface of the sharpening stone, which is especially useful when constant use has worn out some parts more than others. Alternatively, you can flatten using rough grit sandpaper, which removes metal filings and also wears the stone out to create a flat, smooth surface.

81. Using the grinder as a buffer (safely)

A grinder is easily the most useful tool in your smithy. Since sanding takes up as much as 70% of the knifemaking process, you will spend a lot of your time at the grinder shaping your steel to make great blades. Well, it turns out that there are even more uses for the grinder. If you turn to the back of your grinder belt and add some buffing compound, you will have created a buffer/sharpener for yourself. You can then shape your steel, polish it, and sharpen the knife all in one station. Just be sure to clean the belt properly after each use and observe basic grinder use safety.

82. Dealing with heat from your grinder

This is a very important tip; the friction generated from your grinder makes your knife heat up pretty quickly. This means that when you grind for very long, the shape of the blade might even change as it heats up. To keep it cool and in the proper shape and temperature all through the grind, dip your blade in water from time to time.

83. Caution for using smaller grinders

The small grinders might help you start off the grinding process, but they can be a real headache to maintain. Not only do the belts wear out much more quickly than bigger ones, they also tend to produce an inconsistent grind as they wear out. Even if you start with a small grinder because of a small budget, upgrade to larger belt grinders as soon as you can. This is the only way to achieve a great, consistent grind on all your knives.

84. The 2" x 72" belt grinder

The 2"x 72" belt grinder is the best large size grinder you will find out there. I highly recommend getting it when you decide that you are finally ready to upgrade. An added advantage to getting this type of belt is that you will also be granted with a huge selection of accompanying belts. This means that it will finally be possible to change the type of belt you use for every step of the grinding process. Not only will your grinds get better, but your pace of production will improve dramatically as well.

85. The 120 grit wheel

The angle grinder is quite versatile as a replacement for the file. You can use various types of wheels to perform different tasks. For example, the 120 grit wheel is very good when taking off large bits of steel in stock removal. An angle grinder is a power tool that will shave off huge pieces of steel in mere seconds. If you are not careful, you could end up messing up your knife design.

Pro tips

86. Take pride in your work

Every part of the knife, from blade to handle, pommel, and guard contributes to the quality of the final product. You must never attempt to assemble a knife with pieces you are not 100% satisfied with. The first and most ruthless critic of your work must be yourself. Place the highest standards for yourself and strive tirelessly to achieve them. This is how professional knifemakers create a market for themselves.

87. A creative use for coffee

As an acidic liquid itself, instant coffee mix is used by many bladesmiths to etch their Damascus blades. It also comes in handy in darkening the blade's etch by developing a patina over it.

88. Protecting your Damascus steel from rust

As one of the most sought-after blades, Damascus knives are quite valuable. However, just like other knives, they are prone to rusting. To avoid this, you should use renaissance wax to protect Damascus from rust. A good idea is to cover it with a thin coat of wax and then wipe it clean before it dries. Not only does the renaissance wax protect the blade from rust, it also keeps it spotlessly clean and allows it to maintain a nice luster.

89. The proper way to sand

The grit size of your sandpaper determines the final texture of your blade surface. With coarser grits, your blade or handle will be left with large scratches. You can then get it to reach the levels of smoothness you desire by increasing the grit size. An efficient way to do this is to use **progressively higher grits** rather than hopping from grit size to grit size. Start with 200-220 grit sandpaper and then take it to at least 400, progressing until the fine-grained 2,000 grit sandpaper.

90. Using scrap steel

Making knives requires the economical use of materials, including the steel you use to make your blades. Recycling and reusing known steel is one of the most cost effective ways to make knives.

A few things to note when using scrap metals include:

Leaf springs are more often than not similar to 5160 steel. You can get very good blades by simply pounding the bar to remove the curve, then following the knifemaking process as usual to come up with as good a knife as any steel bar can give you. If the piece of leaf spring you have is too large, you can simply cut it down to the size of a normal steel bar to make it easier to grind on.

Another trick is to repurpose your old, worn-out files into blades. You see, most of the good quality files, especially those from Nicholson, are made of 1095 steel. With files, you just grind the surface to remove the ridges, then follow the stock removal process to come up with a knife that you will find keeps a sharp edge even longer than those made from other types of material.

Never use unknown steel. The amount of fuel it would cost you to heat treat unknown steel will be greater than the cost of buying known stock steel.

91. Achieving the flattest flats

Granite countertops make for the best flattening surfaces. You can fashion one from granite sink cutouts from your local granite countertop fabricator. They usually have dumpsters overflowing with good-sized, thick and FLAT pieces of granite and will probably be more than happy to let you cart away as many as you like.

You should then set it up in your workshop and shine a light from the back to check on the flatness of your blades. To make your blade or handle flat, you simply attach a piece of sandpaper to the slab with spray adhesive and hand sand. You will have a dead flat handle or blade in less than a minute by scrubbing it through the sandpaper in a figure 8 motion. The granite slab generates better flats than a belt grinder.

92. The fit and finish are very important

In the world of professional bladesmithing, the finish you put on your knife counts for a lot. You must pay attention to detail if you are to stand a chance of producing knives that your customers will not mind paying a high price for.

The fit means that every part of blade fits perfectly with each other. When forging a knife, you should start by thinking about it in use, then work from there to produce the blade.

The finish is what makes a bladesmith's knife stand out from the millions of mass-produced knives in convenience stalls everywhere. A personal touch, an extra mile, a distinction that makes the user take pride in their knife and care for it - this is what you must always aim for.

When you take care of both the fit and finish, your blades will fetch a higher price and appreciation.

93. Testing your blades in advance

Testing the knives out will allow you to discover any weaknesses and correct them if you can. If you cannot correct them, you can then put the defects in your diary and avoid the same mistakes in the future.

Testing for sharpness

The test for the sharpness of a knife is to cut through a paper towel. If the knife goes through without tearing it, you are good to go.

Shoulder drop test

Drop the knife straight down into concrete floor from shoulder length. If the tip does not bend or chip, you are good to go. If it does any of these things, your hardening and/or tempering processes may be flawed.

Test for toughness of the blade

Hit the sharp edge of the anvil with moderate force with the knife. If the sharp edge chips, your knife was not properly tempered.

Edge Holding test

Whack through a 2- inch x 4-inch piece of wood a few times. Then try to shave off your arm hair with the knife. If it can still shave, the edge is good to go.

94. Follow in the footsteps of the greats

When designing your knife, it helps to compare the design of your blade against the designs of master bladesmiths. By comparing the designs and the results of the tests described above, you will be able to come to figure out the things that work and those that don't. When starting, you can achieve much by making blades that resemble those of your role models. As you improve, you can tweak the designs and add a personal touch.

Similarly, participating in bladesmithing forums, attending bladesmithing conventions, and interacting with fellow bladesmithing will help you keep up with trends, learn new things, and keep the fire of your passion burning even stronger. You will find a lot of inspiration from these spaces.

95. Sanding like a pro

The best way to sand steel is to do every successive rounds from the opposing direction. This enables you to see the scratches from the previous round of sanding much better so that you can tell when they are gone.

96. Pay attention to the feel

Most bladesmiths overlook the feel of their knives. But when a customer picks up one of your knives, they decide to buy or forego it depending on how it looks and how it balances on the hand. A prick from a handle that you were not diligent enough with while sanding, poor balance between blade and handle, or bad beveling could make a customer decide against buying.

97. Bringing it all together

After pouring all your energy into a blade, you want to receive its worth when the customer finally picks it up. Since it will take you a day at the very least to make a really good knife, the price at which you sell it matters. The secret to the knife selling at $3000 is not only the materials, it's using but also its layout, how the design flows, and the finish; integral bolster, how well the guard fits, whether its made of Damascus. If you want to play in that league, ensure that your knife reaches the same quality.

98. A sketch goes a long way

The first place your knife takes shape is in your own mind. You then transfer your idea onto paper before going to the forge or grinder. You must get your knife design properly drawn out on a piece of paper. It will act as your compass all through the knifemaking process. While making a knife through stock removal, a great idea is to cut out the sketch of your knife and paste it onto the piece of steel you'll be making your knife from.

The final product is a direct result of the sketch you make at the beginning so make it a good one.

99. Pro design tip

As a bladesmith, one practical tip I can give you as far as designing your knife goes is to make sure that the lines of your knife design flow. Ensure that your handle proportion balances the blade out.

By creating a visually balanced design sketch, you will get a pretty well-balanced knife at the end.

100. Every action should lead to the next

Bladesmithing requires ironclad focus and concentration in every process, but this does not mean that you shouldn't think ahead. When you make your sketch, it always pays to think through the whole knife making process and outline the various procedures step by step.

When you start, it pays to always keep the next step in mind before proceeding. When you are done with the step, stop and think, and then proceed. This way, every process leads to the next one, the blade makes an allowance for the bolster, which bonds perfectly with the blade and everything works out without a hitch.

101.   Keep everything tidy

The last thing you should do in the knifemaking process is cleaning up after yourself. This allows you to take a step back and analyze your work and also to plan for the next project. Personally, I like to clean up my workshop after every step of the knifemaking process. It helps me get clarity on what to do next and keeps my workspace tidy and safe.

102.   Make your knives in batches

While it is not possible to mass-produce knives through forging or stock removal, you can speed up the process substantially by using a trick I have learned and will now share with you.

Try making knives in batches of 5. This requires a higher level of mastery, coordination, and focus, but it is the only way you can speed up, without sacrificing quality.

103.   Quenching a Japanese katana

The Japanese katana is one of the most difficult blades you will ever make. From the materials, the tools, and the especially the processes of making this majestic sword of old, you will have to bring your A-game. But the Japanese katana is really made at the process of quenching when it is coated with clay and dipped in water.

If and when you decide to make this magnificent blade, you must have the right quench tank because it is dipped sideways rather than downward into the tank. Also, to prepare it for the characteristic bend, you must remember to coat it unevenly with the Hamon clay. The edge should be coated with a thinner layer to facilitate the edge hardening and contraction, with the end hardening more and the rest of the blade curving downward to facilitate the extra hardness.

104.   Take your time

Every blade deserves to be forged to be the best possible knife it can be. The good thing with forging is that you can always take your blade back to the flame and beat it down to your desired shape or size. This is not something you should do in a hurry, so take your time with every blade you forge. If something does not look right, put it down and take a breather. Wait until it feels right. It is better to produce one masterfully done knife than ten shoddy blades. You can also make more money from selling the former than the latter.

Conclusion

So there you have it; starting with the basic tools with which you can kick start your knifemaking career and concluding with the quenching of a Japanese katana-arguably the most sophisticated blade you could ever endeavor to make. And in between, 101 secrets to bladesmithing that will help you master the whole process and answer every burning question you may have. In conclusion:

The most basic tools with which you can launch your bladesmithing career include a file, hacksaw, drill, and sandpaper. You are encouraged to make two or three blades with these basic tools before buying the more expensive and specialized equipment. Other issues discussed in the first chapter include posture, basic safety, and the right mindset for success in bladesmithing. The golden rule; expect to, embrace, and learn from your mistakes.

On tools, the most basic is a good bench vise. It allows you to hold down your blades as you work on them and can improve the quality of your work a great deal. Moreover, it requires little prior skill to operate. But, obviously, there is only so much that you can do with the basic tools. As you grow your skills and improve the quality of your blades, you will need more and more advanced tools and equipment. Some of the more advanced tools include Scotch-Brite wheels for sanding, Farriers rasps for filing, and an air compressor for the more advanced grinding works. The rule to buying tools is that you must only buy those tools that you use enough to justify the investment.

In any workshop, safety is just as important as the work you do there. And with the sharp tools, electrical equipment, and intense heating found in smithies, safety is even more important for a bladesmith. The first rule is that your safety comes first at all times. Wear your tough leather apron at all times, your respirator while quenching to avoid poisonous fumes, and tuck in long hair, beard, and clothes. However, safety is as much the things you wear as those you don't. For example, while operating spinning tools, gloves are a no-no.

Keep everything perfectly organized. Use the 5s methodology to sort, set in order, shine, standardize, and sustain your workshop in the best possible working condition.

And chapter 4, we learned about the foremost material for making knives- steel. The best steel to start working with is the 1084 because it is much easier to heat treat and work into blades. As your skill level increases you can advance to O1 steel. A patina protects blades from rust and Evapo-Rust cleans rusted blades.

A knife is not complete until you have made a handle to allow the user to brandish it. You can make a handle from wood, metal, fabric, and leather among others. Proper knifemaking requires that you drill holes properly, sand everything well, and attach the handle cautiously so as not to blow out the handle material.

As you go professional and start making knives for sale, you must know how to create the ultimate fit and finish, test the quality of your blades, and increase your productivity by making your knives in batches. And to tie it all together, ensure that you engage your fellow bladesmiths in forums for more practical tips and input on every step of the process.

Leave A Review?

Throughout the process of writing this book, I have tried to put down as much value and knowledge for the reader as possible. Some things I knew and practice, some others I spent time to research. I hope you found this book to be of benefit to you!

If you liked the book, would you consider leaving a quick review for it? I would be grateful to you for letting other people know that you like it.

Yours Sincerely,

Wes Sander